Designing the Future: A Student's Approach to Medical Devices

Xenia Ivanova

TABLE OF CONTENTS

Chapter 1: Introduction to Medical Device Design

The Importance of Medical Devices

As students studying biomedical engineering, it is crucial to understand the immense importance of medical devices in the healthcare industry. Medical devices play a pivotal role in improving patient outcomes, enhancing the quality of life, and revolutionizing the way healthcare is delivered. In this subchapter, we will delve into the significance of medical devices and how they contribute to designing a better future.

First and foremost, medical devices are essential tools used by healthcare professionals to diagnose, monitor, and treat various medical conditions. From simple devices like thermometers and blood pressure monitors to advanced imaging technologies such as MRI and CT scanners, these instruments enable accurate diagnosis and effective treatment planning. Medical devices empower healthcare providers to make informed decisions, leading to improved patient care and better treatment outcomes.

Moreover, medical devices are instrumental in preventing and managing chronic diseases. Devices like insulin pumps, pacemakers, and continuous glucose monitors are lifelines for patients with diabetes and cardiovascular diseases. These devices not only aid in maintaining stable health but also allow patients to lead a more active and fulfilling life. Medical devices are designed to be user-friendly, ensuring that patients can manage their conditions independently, fostering self-care and reducing healthcare costs.

In addition to their clinical significance, medical devices also drive innovation and advancements in the field of biomedical engineering. As students, we have the opportunity to contribute to the development of cutting-edge medical technologies that can save lives and revolutionize healthcare. From designing prosthetic limbs with advanced functionality to developing robotic surgical systems that enhance precision, the possibilities are endless. Medical devices have the potential to bridge the gap between engineering and medicine, providing interdisciplinary solutions to complex healthcare challenges.

Furthermore, medical devices are not limited to hospitals and clinics; they are increasingly being used in remote and resource-limited settings. Portable diagnostic devices, telemedicine tools, and wearable sensors enable healthcare delivery in underserved areas, ensuring that everyone has access to quality healthcare. By designing medical devices that are affordable, portable, and easy to use, we can contribute to global health equity and improve healthcare access worldwide.

In conclusion, the importance of medical devices cannot be overstated. They are indispensable in healthcare delivery, facilitating accurate diagnosis, effective treatment, and chronic disease management. Medical devices also offer immense opportunities for innovation and interdisciplinary collaboration. As students of biomedical engineering, we have the privilege and responsibility to design the future of medical devices that will shape the healthcare landscape and improve the lives of millions.

Overview of the Design Process

The design process is a fundamental aspect of creating medical devices in the field of biomedical engineering. It encompasses a systematic approach to problem-solving and innovation, ensuring that the end product meets the needs of patients and healthcare professionals. In this subchapter, we will delve into the various stages of the design process and the importance of each step.

The first stage of the design process is identifying the problem or need. This involves conducting thorough research to understand the challenges faced by patients or healthcare providers. By gaining insights into their requirements, students can begin to brainstorm potential solutions.

Once the problem has been identified, the next step is to gather and analyze data. This includes studying existing devices, conducting surveys, and interviewing experts in the field. By collecting and synthesizing this information, students can gain a deeper understanding of the problem and potential solutions.

The third stage is concept development. This is where ideas are generated and refined. Students can use techniques like brainstorming and mind mapping to explore different possibilities. It is essential to consider factors like feasibility, safety, and effectiveness during this phase.

After concept development, the next stage is prototyping. This involves creating physical or virtual models of the proposed device. Prototypes allow students to test and validate their ideas, making necessary adjustments along the way. It is crucial to iterate on prototypes to refine the design and ensure it meets the desired objectives.

Once a prototype has been finalized, the design moves into the manufacturing stage. This involves scaling up the production process and ensuring that the device can be manufactured reliably and cost-effectively. It is essential to consider manufacturing constraints and regulations during this phase.

The final stage of the design process is testing and evaluation. This involves conducting rigorous trials to verify the device's safety, efficacy, and usability. Students should consider user feedback and iterate on the design based on the test results.

Throughout the design process, it is crucial to consider ethical, social, and environmental factors. Students must ensure that their designs are inclusive, accessible, and sustainable.

In conclusion, the design process is a systematic approach to creating medical devices in the field of biomedical engineering. By following the stages of problem identification, data gathering, concept development, prototyping, manufacturing, and testing, students can ensure that their designs meet the needs of patients and healthcare professionals. It is essential to consider various factors like feasibility, safety, and effectiveness throughout the design process to create innovative and impactful medical devices.

Ethical Considerations in Medical Device Design

Introduction:
In the ever-evolving field of biomedical engineering, the design and development of medical devices play a crucial role in revolutionizing healthcare. However, it is important for students in this field to understand the ethical considerations that should be taken into account during the design process. This subchapter aims to provide a comprehensive overview of the ethical considerations in medical device design, equipping students with the knowledge and understanding necessary to create innovative and ethically responsible solutions.

Patient   Safety:
One of the primary ethical considerations in medical device design is patient safety. Students must prioritize the well-being of the patients who will use the devices they create. This involves conducting thorough risk assessments, adhering to strict regulatory standards, and implementing rigorous quality assurance measures throughout the design and manufacturing processes.

Informed   Consent:
Informed consent is another crucial ethical aspect that students must consider. It is essential that patients fully understand the risks, benefits, and potential alternatives associated with a medical device. Students should learn to communicate effectively with patients, ensuring they make informed decisions about their treatment options.

Equitable   Access:
Ensuring equitable access to medical devices is a critical ethical

consideration. Students should strive to create devices that are affordable, accessible, and culturally sensitive. By considering the economic and social factors that impact healthcare access, students can design devices that benefit all segments of society.

Data Privacy and Security:
With the increasing integration of technology in medical devices, data privacy and security have become significant ethical concerns. Students must understand the importance of protecting patient data and designing systems that safeguard confidentiality. It is their responsibility to ensure that devices are equipped with appropriate security measures to prevent unauthorized access and potential breaches.

Sustainability:
In recent years, the concept of sustainability has gained prominence in medical device design. Students should consider the environmental impact of their devices throughout their lifecycle, from raw material sourcing to disposal. They should explore innovative materials and manufacturing processes that minimize waste and energy consumption, promoting a greener and more sustainable future.

Conclusion:
Ethical considerations play a pivotal role in the design of medical devices, and it is essential for students of biomedical engineering to be well-versed in these principles. By prioritizing patient safety, informed consent, equitable access, data privacy, and sustainability, students can contribute to the development of medical devices that not only improve healthcare outcomes but also uphold ethical standards. This subchapter serves as a foundation for students, enabling them to make

informed design decisions and create a positive impact on the field of biomedical engineering.

Chapter 2: Understanding User Needs

Conducting User Research

User research is a crucial step in the process of designing medical devices. It involves gathering insights and feedback from potential users to better understand their needs, preferences, and challenges. By conducting user research, biomedical engineers can ensure that their designs are user-centered, safe, and effective.

There are several methods and techniques that students in the field of biomedical engineering can use to conduct user research. One of the most common approaches is conducting interviews with potential users. By asking open-ended questions and actively listening to their responses, students can gain valuable insights into the users' experiences, expectations, and pain points. This qualitative research method helps identify design opportunities and areas for improvement.

Observation is another powerful tool for user research. Students can observe potential users in their natural environments to understand how they interact with existing medical devices or cope with certain health conditions. This allows engineers to identify usability issues and design solutions that better meet the users' needs.

In addition to interviews and observation, surveys and questionnaires can also be used to collect quantitative data on user preferences, satisfaction levels, and usability ratings. By analyzing this data, students can identify patterns and trends, which can guide the design process and inform decision-making.

Another valuable technique students can employ is prototyping. Building low-fidelity prototypes allows them to gather feedback early in the design process. Users can interact with the prototypes and provide feedback on their functionality, usability, and aesthetics. This iterative approach enables students to refine their designs based on user input, resulting in more user-friendly and effective medical devices.

Furthermore, involving potential users in the design process through co-creation sessions or focus groups can foster collaboration. By actively engaging users, students can gain a deeper understanding of their needs and preferences and ensure that the final product meets their expectations.

Overall, conducting user research is essential for biomedical engineering students when designing medical devices. It helps them understand the needs of potential users, identify design opportunities, and create devices that are safe, effective, and user-centered. By utilizing a variety of research methods and techniques, students can gather valuable insights, refine their designs, and ultimately make a positive impact on healthcare.

Defining User Personas

In the world of biomedical engineering, understanding the needs and preferences of users is crucial for designing effective medical devices. This is where user personas come into play. User personas are fictional representations of different types of users, based on research and data, that help designers empathize with the end users and make informed design decisions. In this subchapter, we will delve into the significance of defining user personas in the context of medical device design.

User personas serve as archetypes that represent the characteristics, behaviors, needs, and goals of specific user groups. By creating these personas, biomedical engineers can gain a better understanding of the diverse range of users they are designing for. For instance, in the field of medical devices, user personas can range from patients with chronic illnesses to healthcare professionals who operate the devices. Each persona represents a distinct user group, and understanding their unique requirements is essential for designing tailored solutions.

To define user personas effectively, thorough research and data analysis are crucial. This involves conducting interviews, surveys, and observations to gather insights about potential users. By identifying patterns and commonalities among users, we can create personas that accurately represent the target audience. For instance, when designing a wearable device for patients with diabetes, personas could include a tech-savvy young adult who requires seamless integration with their smartphone, as well as an elderly patient who values simplicity and ease of use.

User personas also help in aligning the design process with user-centered principles. By referring to the personas throughout the design process, biomedical engineering students can ensure that the end product caters to the specific needs of the target users. This approach fosters innovation and enhances the overall user experience, leading to more successful and impactful medical devices.

Furthermore, user personas can aid in effective communication and collaboration among multidisciplinary teams. When designers, engineers, and healthcare professionals have a shared understanding of the user personas, it becomes easier to communicate ideas, share insights, and work towards a common goal. This collaboration is vital in ensuring that the final product meets the needs of all stakeholders involved.

In conclusion, defining user personas is an essential step in the design process of medical devices. By understanding the charactcristics, behaviors, needs, and goals of different user groups, biomedical engineering students can create devices that are user-centered and impactful. User personas not only help in empathizing with the end users but also facilitate effective communication and collaboration among multidisciplinary teams. By embracing user personas, students can pave the way for designing the future of medical devices that are both innovative and cater to the unique needs of patients and healthcare professionals.

Identifying User Requirements

In the field of biomedical engineering, the process of designing and developing medical devices is a critical one. As aspiring biomedical engineers, it is essential for us to understand the importance of identifying user requirements before embarking on any design project. This subchapter aims to provide students in the field of biomedical engineering with insights into the process of identifying user requirements for medical devices.

User requirements refer to the specific needs and expectations of the end-users or patients who will be utilizing the medical devices. These requirements play a pivotal role in guiding the design and development process, as they ensure that the devices created are safe, effective, and user-friendly.

The first step in identifying user requirements is to conduct thorough research. Students must familiarize themselves with the medical condition or problem that the device aims to address. This involves studying existing literature, consulting healthcare professionals, and even interacting with patients to gain insights into their experiences and challenges.

Once a solid foundation of knowledge has been established, the next step is to create user profiles. Students should identify the different types of users who will interact with the device, such as patients, doctors, or caregivers. By understanding the unique needs and constraints of each user group, engineers can tailor the design to meet their specific requirements.

To ensure a comprehensive understanding of user requirements, it is crucial to engage in user testing and feedback sessions. This allows engineers to observe how users interact with prototypes or mock-ups of the device and gather valuable insights. Students can organize focus groups or conduct surveys to obtain feedback directly from the target audience. This iterative process helps refine the design and ensures that it aligns with user expectations.

Furthermore, it is important to consider regulatory standards and guidelines while identifying user requirements. Medical devices must comply with strict regulations to ensure patient safety. Students should be aware of these standards and integrate them into the design process from the early stages.

By thoroughly identifying user requirements, students in the field of biomedical engineering can develop medical devices that cater to the specific needs of patients and healthcare professionals. This process not only enhances the effectiveness and usability of the device but also contributes to the overall improvement of healthcare outcomes.

Chapter 3: Ideation and Concept Generation

Brainstorming Techniques

In the field of biomedical engineering, creativity and innovation play a vital role in designing medical devices that can revolutionize healthcare. Brainstorming is a powerful technique that helps unleash your creative potential and generate new ideas. This subchapter explores various brainstorming techniques that can be employed by students in the field of biomedical engineering to design the future of medical devices.

1. Traditional Brainstorming: This widely-used technique involves generating a large number of ideas within a short span of time. Students gather in a comfortable and open environment, where no idea is considered too outrageous or impractical. The focus is on quantity rather than quality, as the aim is to encourage free thinking and generate a pool of ideas that can be refined later.

2. Mind Mapping: Mind mapping is a visual brainstorming technique that helps students organize their thoughts and explore connections between different concepts. It involves creating a diagram or flowchart that branches out from a central idea. Students can use colors, symbols, and keywords to visually represent their ideas and develop them further.

3. Reverse Brainstorming: This technique involves flipping the problem on its head and brainstorming ways to create the opposite effect. For example, if the problem is to improve the accuracy of a diagnostic tool, students can brainstorm ideas on how to make it less

accurate. By reversing the problem, students can uncover unique perspectives and innovative solutions.

4. SCAMPER Technique: SCAMPER stands for Substitute, Combine, Adapt, Modify, Put to another use, Eliminate, and Reverse. This technique encourages students to think outside the box by asking questions related to each category. For example, students can ask themselves how they can modify an existing medical device to make it more user-friendly or combine two different devices to create a more efficient solution.

5. Random Stimulus Technique: This technique involves randomly selecting a word, image, or object and using it as a stimulus for generating ideas. Students can use online random word generators or physical objects to inspire their brainstorming sessions. The randomness of the stimulus forces students to think creatively and make unexpected connections.

By incorporating these brainstorming techniques into their design process, students in the field of biomedical engineering can tap into their creative potential and develop innovative solutions for medical devices. Remember, brainstorming is a skill that can be honed with practice, so don't be afraid to experiment and think outside the box. The future of medical devices lies in the hands of creative and imaginative students like you.

Sketching and Prototyping

In the fast-paced world of biomedical engineering, sketching and prototyping play a vital role in the design process of medical devices. This subchapter will delve into the importance of sketching and prototyping and how they contribute to the development of innovative solutions in the field.

Sketching is a fundamental skill that allows biomedical engineers to visually communicate their ideas and concepts. Whether it's brainstorming new concepts or refining existing designs, sketching enables students to quickly capture and explore different possibilities. By putting ideas on paper, students can visualize their thoughts, iterate on designs, and collaborate with their peers. Sketching also helps in identifying potential design flaws and improving the functionality and usability of medical devices.

Prototyping is another critical step in the design process. It involves creating physical models or mock-ups of the envisioned medical device. Prototypes allow students to test and evaluate the functionality, ergonomics, and aesthetics of their designs. By creating physical representations, students can gather valuable feedback, identify areas for improvement, and make necessary modifications before moving forward with the final product. Prototyping also helps in understanding the manufacturing requirements and feasibility of the design, ensuring that the medical device can be produced efficiently and cost-effectively.

Both sketching and prototyping are iterative processes that encourage students to think critically and creatively. They provide a platform for experimentation, exploration, and problem-solving. Through sketching and prototyping, students can bridge the gap between theory and practice, transforming their ideas into tangible solutions.

To enhance sketching and prototyping skills, students can utilize various tools and techniques. From traditional pen and paper to digital software, there are numerous resources available to streamline the design process. Additionally, workshops and courses on sketching and prototyping can provide hands-on experience and guidance in effectively translating ideas into reality.

In conclusion, sketching and prototyping are indispensable tools for biomedical engineering students. They enable the visualization of ideas, facilitate collaboration, and help in the creation of efficient and user-friendly medical devices. By mastering these skills, students can bring their innovative concepts to life and contribute to the advancement of healthcare technology.

Design Validation and Testing

In the realm of biomedical engineering, design validation and testing play a crucial role in ensuring the safety and efficacy of medical devices. As aspiring biomedical engineers, it is essential to understand the importance of this stage in the design process.

Design validation refers to the process of evaluating a medical device to ensure that it meets the intended use and user needs. This phase involves conducting tests, experiments, and simulations to verify and validate the design's performance. The goal is to identify any potential issues or shortcomings before the device is manufactured and released for use in the real world.

Testing is a vital component of design validation, as it allows engineers to assess the device's functionality, reliability, durability, and safety. There are various types of tests that biomedical engineers employ, including mechanical testing, electrical testing, and biological testing. Mechanical testing involves subjecting the device to various physical stresses, such as compression, tension, and torsion, to evaluate its structural integrity. Electrical testing, on the other hand, examines the electrical components and circuits of the device to ensure proper functioning. Lastly, biological testing assesses the device's interaction with living tissues and organisms to determine its biocompatibility and potential risks.

To conduct these tests, biomedical engineers utilize advanced tools and techniques such as computer simulations, prototype fabrication, and data analysis software. Computer simulations allow engineers to virtually test and optimize the design before creating physical prototypes, saving both time and resources. Prototypes are then fabricated to conduct real-world tests, enabling engineers to gather valuable data and make necessary improvements.

Design validation and testing are iterative processes that involve multiple rounds of testing and refinement. It is essential to document and analyze the results at each stage, as this information helps in identifying design flaws and making informed design decisions. The ultimate goal is to create a safe, reliable, and effective medical device that meets the needs of patients and healthcare professionals.

As students pursuing a career in biomedical engineering, it is crucial to develop a strong understanding of design validation and testing. By mastering these concepts, we can contribute to the development of innovative and life-changing medical devices that have a positive impact on the healthcare industry.

Chapter 4: Designing for Human Factors

Ergonomics and Usability

Ergonomics and Usability: Enhancing Medical Device Design for a User-Centered Approach

Introduction:
In the field of biomedical engineering, designing medical devices that meet the needs of both healthcare professionals and patients is of utmost importance. One key aspect that cannot be overlooked is the integration of ergonomics and usability in medical device design. This subchapter aims to shed light on the significance of ergonomics and usability and how they can be effectively incorporated to enhance the functionality and user experience of medical devices.

Understanding   Ergonomics:
Ergonomics is the science of designing products and systems that effectively fit the physical and cognitive capabilities of users. In the context of medical devices, this involves creating designs that optimize user comfort, minimize the risk of injury, and enhance overall efficiency. Students in biomedical engineering must recognize the importance of ergonomics in order to develop medical devices that are safe, intuitive, and user-friendly.

Usability   in   Medical   Device   Design:
Usability refers to the ease with which users can interact with a device to achieve their desired goals. In the medical field, usability plays a crucial role in ensuring that devices are efficient, reliable, and intuitive for both healthcare professionals and patients. Students must consider

factors such as device navigation, information presentation, and user feedback to create intuitive interfaces and streamline the overall user experience.

Human-Centered Design Approach: An effective way to incorporate ergonomics and usability in medical device design is by adopting a human-centered design approach. This approach emphasizes involving end-users throughout the design process, allowing for their input and feedback. By understanding the needs, capabilities, and limitations of users, students can create medical devices that truly address their requirements, resulting in improved patient outcomes and user satisfaction.

Considerations for Ergonomic Design: When designing medical devices, students should consider various ergonomic factors, such as device size and weight, ease of handling, and the positioning of controls and displays. Additionally, considering the physical attributes of the target user population, such as age, mobility, and dexterity, is crucial in achieving optimal ergonomics. By addressing these factors, students can ensure that medical devices are comfortable and easy to use for a wide range of users.

Enhancing Usability through User-Centered Design: To enhance usability, students must focus on designing intuitive interfaces, clear instructions, and meaningful feedback systems. Implementing user-centered design principles, such as conducting usability tests and gathering user feedback, allows for iterative improvements and ensures that medical devices meet the specific needs of users. By optimizing usability, students can enhance the overall user experience, leading to increased device adoption and user satisfaction.

Conclusion:

Incorporating ergonomics and usability into medical device design is crucial for creating devices that are safe, efficient, and user-friendly. By considering the physical and cognitive capabilities of users, students in biomedical engineering can create medical devices that address the specific needs of both healthcare professionals and patients. By adopting a human-centered design approach and focusing on ergonomic design and usability enhancements, students can play a vital role in shaping the future of medical devices, ultimately improving patient outcomes and healthcare experiences.

Human-Machine Interaction

In the world of biomedical engineering, the concept of Human-Machine Interaction (HMI) is of utmost importance. HMI refers to the interaction between humans and machines, specifically in the context of medical devices. As students in the field of biomedical engineering, understanding and mastering HMI is crucial for designing the future of medical devices.

HMI plays a vital role in ensuring the effectiveness, safety, and usability of medical devices. It encompasses various elements such as user interface design, ergonomics, and the overall user experience. The goal is to create devices that seamlessly integrate with human capabilities and needs, enhancing patient care and improving medical outcomes.

When designing medical devices, it is essential to consider the needs and abilities of the end-users, who could be patients, doctors, nurses, or other healthcare professionals. A deep understanding of human behavior, cognitive processes, and physical limitations is necessary to develop devices that are intuitive and easy to use. By applying principles of psychology, ergonomics, and human factors engineering, biomedical engineers can optimize HMI and ensure that devices are user-friendly and efficient.

One aspect of HMI that deserves special attention is the user interface design. The interface is the bridge between the user and the device, allowing for communication and control. It should be designed in a way that minimizes the learning curve, reduces errors, and maximizes efficiency. Intuitive icons, clear instructions, and logical workflows are some of the key elements in creating an effective user interface.

Another consideration in HMI is the physical interaction between humans and machines. Medical devices should be designed with ergonomics in mind, ensuring that they are comfortable to hold, operate, and interact with. Factors such as size, weight, and tactile feedback should be carefully considered to minimize user fatigue and maximize usability.

Furthermore, the user experience should be taken into account when designing medical devices. A positive user experience not only increases user satisfaction but also promotes compliance and adherence to treatment. By focusing on aspects such as aesthetics, ease of use, and emotional engagement, biomedical engineers can create devices that patients will feel comfortable using, enhancing their overall healthcare experience.

In conclusion, Human-Machine Interaction is a critical aspect of designing medical devices in the field of biomedical engineering. By understanding the needs and abilities of the end-users, optimizing user interface design, considering ergonomics, and prioritizing the user experience, biomedical engineers can create innovative and effective devices that have a positive impact on patient care. As students, mastering HMI will be instrumental in shaping the future of medical devices and revolutionizing healthcare.

Considering Accessibility and Inclusivity

In the field of biomedical engineering, the development of medical devices plays a crucial role in improving the quality of life for individuals with various medical conditions. However, it is essential to consider accessibility and inclusivity when designing these devices to ensure that they can be effectively used by all individuals, regardless of their physical abilities or limitations.

Accessibility refers to the ease with which a person can interact with a device or environment. When designing medical devices, it is important to make them accessible to individuals with disabilities or impairments. This can be achieved by incorporating features such as large buttons, tactile feedback, and adjustable settings to accommodate different user needs. Additionally, considering the needs of individuals with visual or hearing impairments is vital, and the inclusion of audio or visual cues can greatly enhance accessibility.

Inclusivity, on the other hand, focuses on designing devices that are inclusive of all individuals, regardless of their age, gender, or cultural background. It is crucial to consider the diversity of the user population and ensure that the device functions effectively for all users. This can be achieved by conducting user research and involving individuals from different backgrounds in the design process. By considering the needs and preferences of diverse users, designers can create devices that are more inclusive and user-friendly.

Moreover, considering accessibility and inclusivity not only benefits individuals with disabilities but also provides advantages to the wider user population. For instance, a medical device that is easy to use and understand benefits everyone, including those with limited technical knowledge or older individuals who may have difficulty adapting to new technologies. By designing with accessibility and inclusivity in mind, biomedical engineers can create devices that are intuitive, user-friendly, and cater to the needs of a diverse user population.

In conclusion, when designing medical devices, it is essential to prioritize accessibility and inclusivity. By considering the needs of individuals with disabilities and ensuring that devices are inclusive of all users, biomedical engineers can create innovative and user-friendly solutions that have a positive impact on the lives of individuals with medical conditions. Through a student's approach to medical device design, we can shape a future where healthcare is accessible to all, regardless of physical abilities or limitations.

Chapter 5: Materials and Manufacturing in Medical Device Design

Material Selection in Medical Devices

In the field of biomedical engineering, the selection of materials for medical devices plays a crucial role in ensuring the safety, reliability, and effectiveness of these devices. The choice of materials not only affects the durability and performance of medical devices but also impacts patient comfort and overall treatment outcomes. This subchapter aims to provide students with an understanding of the importance of material selection in the design and development of medical devices.

One of the primary considerations in material selection is biocompatibility. Medical devices come into contact with the human body, and therefore, the materials used must be non-toxic, non-allergenic, and not induce any adverse reactions. Students will learn about the various biocompatibility tests and standards used to evaluate materials for medical devices. They will also explore the different classes of medical devices and the specific material requirements for each class.

Another critical aspect of material selection is mechanical properties. Students will gain insights into the mechanical characteristics of materials such as strength, stiffness, and flexibility, and how these properties impact the functionality and performance of medical devices. They will understand the importance of selecting materials with suitable mechanical properties to ensure the device can withstand the required loads and stresses during use.

The subchapter will delve into the unique challenges associated with material selection for implantable medical devices. Students will explore the different types of implants and the specific material requirements for each type. They will learn about the corrosion resistance, wear resistance, and fatigue properties of implant materials to ensure long-term stability and functionality within the human body.

Furthermore, students will be introduced to the advancements in material science and technology that have revolutionized the field of medical devices. They will discover emerging materials such as shape memory alloys, biodegradable polymers, and nanomaterials, and their potential applications in medical device design. The subchapter will encourage students to consider the future implications of material selection in medical devices and the role they can play in driving innovation.

Overall, this subchapter on material selection in medical devices aims to equip students with the knowledge and skills necessary to make informed decisions when designing and developing medical devices. By understanding the significance of biocompatibility, mechanical properties, and advancements in material science, students will be well-prepared to contribute to the future of biomedical engineering and make a positive impact on patient care.

Manufacturing Processes and Techniques

In the field of biomedical engineering, the development and production of medical devices play a vital role in improving healthcare outcomes and enhancing the quality of life for patients. This subchapter, "Manufacturing Processes and Techniques," aims to provide students in the niche of biomedical engineering with a comprehensive understanding of the various methods and techniques used in the manufacturing of medical devices.

1.   Introduction   to   Manufacturing   Processes: This section will introduce students to the fundamental concepts of manufacturing processes and their importance in the biomedical engineering field. It will highlight the significance of efficient manufacturing techniques in ensuring the production of safe, reliable, and cost-effective medical devices.

2.   Traditional   Manufacturing   Techniques: Here, students will explore the conventional manufacturing techniques commonly employed in the production of medical devices. This will include processes such as casting, forging, machining, and molding. The section will delve into the advantages, limitations, and suitable applications of each technique.

3.   Advanced   Manufacturing   Technologies: In this section, students will be introduced to cutting-edge manufacturing technologies transforming the biomedical engineering landscape. Topics covered may include additive manufacturing (3D printing), laser cutting, electrospinning, and microfabrication techniques. The discussion will emphasize the potential benefits of

these advanced techniques, such as increased customization, reduced costs, and improved functionality.

4.   Quality   Control   and   Regulatory   Compliance: Manufacturing medical devices requires adherence to strict quality control standards and regulatory compliance. This section will familiarize students with the importance of quality control in ensuring the safety and effectiveness of medical devices. It will also touch upon relevant regulatory bodies and their guidelines, such as the FDA (Food and Drug Administration) in the United States.

5.   Material   Selection   and   Biocompatibility: Choosing suitable materials for medical devices is crucial in ensuring biocompatibility and long-term functionality. This section will highlight the considerations and challenges involved in material selection, including biocompatibility testing, sterilization, and the impact of material properties on device performance.

6.   Case   Studies   and   Industry   Examples: To provide practical insights, this section will present case studies and real-world examples of successful manufacturing processes in the biomedical engineering field. Students will gain valuable knowledge about the challenges faced by industry professionals and the innovative solutions they have implemented.

By the end of this subchapter, students will have a comprehensive understanding of the manufacturing processes and techniques employed in the biomedical engineering field. This knowledge will be valuable for their future careers, enabling them to contribute to the development of safe, efficient, and innovative medical devices.

Quality Assurance and Regulatory Compliance

In the field of biomedical engineering, ensuring the safety and efficacy of medical devices is of utmost importance. This subchapter will delve into the critical aspects of quality assurance and regulatory compliance that students in the field need to be familiar with to design innovative and reliable medical devices.

Quality assurance refers to the systematic process of monitoring and evaluating the various stages of product development to ensure that the final product meets the required standards. Students must understand the importance of quality assurance and how it impacts the performance and reliability of medical devices. This subchapter will provide an overview of the quality assurance process, including quality control techniques, validation, and verification methods. Students will learn about the importance of documentation, risk assessment, and traceability in maintaining quality throughout the design and manufacturing process.

Regulatory compliance is another vital aspect that students must consider when designing medical devices. This subchapter will introduce students to the various regulatory bodies and standards that govern the medical device industry. It will provide an overview of the regulatory landscape, including organizations such as the Food and Drug Administration (FDA) in the United States, the European Medicines Agency (EMA) in Europe, and the International Organization for Standardization (ISO) globally. Students will gain an understanding of the regulatory requirements, including pre-market approval, post-market surveillance, and labeling and reporting obligations.

Additionally, this subchapter will delve into specific regulations and standards that pertain to different types of medical devices, such as implantable devices, diagnostic equipment, and software. Students will learn about the design controls, risk management, and post-market surveillance requirements specific to each category of medical devices. Case studies and examples of regulatory compliance challenges will be provided to enhance students' understanding of the real-world implications of regulatory requirements.

By the end of this subchapter, students will have a comprehensive understanding of the importance of quality assurance and regulatory compliance in the design and development of medical devices. They will be equipped with the knowledge and tools necessary to navigate the regulatory landscape and ensure that their designs meet the highest standards of safety and effectiveness. Whether pursuing a career in research, development, or manufacturing, this subchapter will provide students with the foundation needed to design the future of medical devices responsibly and ethically.

Chapter 6: Designing for Safety and Risk Management

Risk Analysis and Assessment

In the field of biomedical engineering, the development and design of medical devices play a critical role in enhancing healthcare delivery and improving patient outcomes. However, it is essential to recognize that these devices also pose potential risks to patients and users if not properly addressed and managed. This subchapter aims to provide students of biomedical engineering with a comprehensive understanding of risk analysis and assessment in the context of medical device design.

Risk analysis refers to the systematic process of identifying, evaluating, and prioritizing potential risks associated with a medical device throughout its lifecycle. It involves assessing the likelihood and severity of harm that could occur, as well as identifying potential hazards and their root causes. By conducting thorough risk analysis, engineers can identify design flaws, manufacturing defects, and other vulnerabilities that may compromise the safety and effectiveness of the device.

One of the fundamental tools for risk analysis is the risk assessment matrix, which allows engineers to evaluate the probability and severity of risks. By assigning numerical values to these factors, engineers can prioritize risks and focus on mitigating those with the highest potential for harm. Additionally, risk assessment involves considering the context of device usage, such as the intended user population, the environment in which it will be used, and the potential consequences of device failure.

To ensure comprehensive risk analysis and assessment, it is crucial to follow established standards and guidelines. The International Organization for Standardization (ISO) has developed several standards specifically addressing risk management in medical devices, such as ISO 14971. Familiarity with these standards is essential for biomedical engineering students to ensure that their designs adhere to best practices and regulatory requirements.

Moreover, risk analysis and assessment should be an ongoing process throughout the device's lifecycle. As new information becomes available, engineers must continuously re-evaluate and update risk assessments to address emerging risks or design flaws. Regular communication and collaboration with healthcare professionals, regulatory bodies, and other stakeholders are vital to gather feedback and ensure that risk mitigation strategies remain effective.

By incorporating risk analysis and assessment into the design process, biomedical engineering students can create safer and more reliable medical devices. Understanding and managing potential risks will not only enhance patient safety but also contribute to the overall success and acceptance of their innovations in the healthcare industry.

Safety Standards and Regulations

In the field of biomedical engineering, the development and design of medical devices require strict adherence to safety standards and regulations. These guidelines exist to ensure the safety and efficacy of medical devices, protecting both patients and healthcare professionals. In this subchapter, we will explore the importance of safety standards and regulations in the design process, providing students with a comprehensive understanding of their role in creating innovative and reliable medical devices.

The field of biomedical engineering encompasses a wide range of medical devices, including diagnostic tools, prosthetics, imaging systems, and therapeutic devices. As students, it is crucial to understand the significance of safety standards and regulations as they directly impact the quality and reliability of these devices. Compliance with these standards is not only a legal requirement but also a moral obligation to prioritize patient safety.

One such standard is the International Organization for Standardization (ISO) 13485, which outlines the quality management system requirements for the design and manufacturing of medical devices. This standard ensures that the design process is systematic and controlled, emphasizing risk management and the traceability of materials and components. By following ISO 13485, students can ensure that their designs meet the highest quality standards, minimizing the potential risks associated with medical devices.

Additionally, regulatory bodies such as the Food and Drug Administration (FDA) in the United States and the European

Medicines Agency (EMA) in Europe play a vital role in ensuring the safety and effectiveness of medical devices. These organizations review and approve medical devices before they can be marketed and sold. Students must be aware of the regulatory requirements specific to their region to navigate the complex process of obtaining regulatory approval.

Understanding safety standards and regulations is not limited to the design phase but extends throughout the entire lifecycle of a medical device. This includes post-market surveillance, which involves monitoring the device's performance, identifying adverse events, and implementing corrective actions when necessary. By actively participating in post-market surveillance, students can contribute to the ongoing improvement and safety of medical devices.

In conclusion, safety standards and regulations form the backbone of the biomedical engineering field. Students must familiarize themselves with these guidelines to design and develop medical devices that prioritize patient safety. Compliance with standards such as ISO 13485 and obtaining regulatory approval from organizations like the FDA and EMA are essential steps in ensuring the reliability and effectiveness of medical devices. By upholding these standards, future biomedical engineers can contribute to the advancement of healthcare and improve the lives of countless individuals.

Mitigating and Managing Risks

In the field of biomedical engineering, designing and developing medical devices is an exciting yet challenging task. The process involves a series of steps, from conceptualization to commercialization, that require careful consideration of potential risks. Mitigating and managing risks is an essential aspect of ensuring the safety, effectiveness, and reliability of medical devices. This subchapter aims to provide students in the field of biomedical engineering with valuable insights into the various strategies and techniques employed to address risks and ensure the success of their medical device designs.

One of the primary steps in mitigating risks is conducting a thorough risk assessment. This involves identifying potential hazards and analyzing their potential consequences. Students will learn about different methods such as Failure Modes and Effects Analysis (FMEA) and Hazard Analysis and Critical Control Points (HACCP), which enable them to identify, evaluate, and prioritize risks. By understanding the potential risks associated with their medical device designs, students can take proactive measures to minimize or eliminate them.

Once risks are identified, students will explore various risk management strategies. Implementing design controls, such as using redundant systems or incorporating fail-safe mechanisms, can help minimize risks and enhance the safety of medical devices. Students will also learn about the importance of adhering to regulatory guidelines and standards, such as ISO 13485 and FDA regulations, which provide a framework for risk management in the medical device industry.

Furthermore, students will gain insights into the importance of usability testing and human factors engineering in risk mitigation. Ensuring that medical devices are intuitive, user-friendly, and error-resistant can significantly reduce the likelihood of user-related errors and adverse events. Students will learn about techniques like task analysis, user feedback, and iterative design processes that enable them to optimize the usability and safety of their medical device designs.

In conclusion, mitigating and managing risks is a critical aspect of designing safe and effective medical devices. This subchapter equips students in the field of biomedical engineering with the necessary knowledge and tools to identify, evaluate, and address potential risks associated with their designs. By implementing proactive risk management strategies and adhering to regulatory guidelines, students can contribute to the development of innovative and reliable medical devices that improve patient outcomes and revolutionize healthcare.

Chapter 7: Design Verification and Validation

Testing and Evaluation Methods

In the field of biomedical engineering, the development of safe and effective medical devices is of utmost importance. To ensure the reliability and functionality of these devices, rigorous testing and evaluation methods are employed throughout the design process. This subchapter will explore the various techniques used by biomedical engineers to test and evaluate medical devices, providing students with a comprehensive understanding of this crucial aspect of their work.

One of the primary testing methods used in biomedical engineering is in vitro testing. This involves conducting experiments in a controlled laboratory environment using simulated biological systems or cell cultures. In vitro testing allows engineers to evaluate the performance of a medical device under specific conditions, providing valuable insights into its functionality and potential limitations.

Another commonly employed method is in vivo testing, which involves conducting experiments on living organisms. This type of testing is often performed on animal models to assess the safety and efficacy of medical devices before they are introduced to human subjects. In vivo testing provides engineers with valuable data on the device's interactions with living tissues, allowing them to identify any potential adverse effects and refine the design accordingly.

In addition to these traditional testing methods, computer simulations and modeling play a significant role in evaluating medical devices. By using advanced software tools, engineers can simulate the behavior of a device under various scenarios, predicting its performance and identifying potential design flaws. Computer simulations help save time and resources by allowing engineers to iterate on the design virtually before moving on to physical prototypes.

Furthermore, human factors testing is an essential aspect of medical device evaluation. This involves assessing the device's usability and user experience through user testing and feedback. By involving potential users in the evaluation process, engineers can identify any design flaws or user interface issues, ensuring that the device is intuitive, efficient, and safe to use.

In conclusion, testing and evaluation methods are critical in the development of medical devices. Biomedical engineers employ a range of techniques, including in vitro and in vivo testing, computer simulations, and human factors testing, to ensure the safety, efficacy, and usability of these devices. By understanding and applying these methods, students in the field of biomedical engineering can contribute to the design and development of innovative and life-saving medical devices.

Usability Testing and Human Factors Validation

As aspiring biomedical engineers, it is crucial for us to understand the significance of usability testing and human factors validation in the design and development of medical devices. In this subchapter, we will delve into the importance of these processes and how they ensure the safety and effectiveness of the devices we create.

Usability testing is a method used to evaluate the ease of use and overall user experience of a medical device. It involves observing and collecting feedback from potential users in order to identify any design flaws or areas for improvement. By conducting usability testing, we can gain valuable insights into how our devices are being perceived and used by the end-users.

Human factors validation, on the other hand, focuses on assessing the compatibility of a medical device with the human operator or user. It involves studying the interaction between the user, the device, and the environment in which it will be used. By considering human factors, such as ergonomics and user behavior, we can design devices that are not only efficient but also safe and intuitive to use.

The significance of usability testing and human factors validation cannot be overstated. These processes enable us to identify and rectify potential risks or errors that may arise during device usage, ensuring patient safety and reducing the likelihood of accidents or adverse events. By involving end-users in the design and testing phases, we can also generate valuable feedback that can help us refine our devices further.

Incorporating usability testing and human factors validation into the design process also adds value to our work from a business perspective. Medical devices that are user-friendly and intuitive to operate are more likely to be adopted by healthcare professionals and patients alike. Additionally, by identifying and addressing design flaws early on, we can save time and resources that would otherwise be wasted on costly redesigns or recalls.

In conclusion, usability testing and human factors validation play a crucial role in the design and development of medical devices. As biomedical engineering students, it is essential for us to understand and implement these processes in our work. By prioritizing user experience and safety, we can contribute to the creation of innovative and effective medical devices that have a positive impact on healthcare delivery.

Clinical Trials and Regulatory Approvals

In the world of biomedical engineering, the development and deployment of medical devices are crucial for improving patient care and enhancing healthcare outcomes. However, before these innovative devices can reach the hands of healthcare professionals and patients, they must undergo rigorous testing and obtain regulatory approvals. This subchapter will delve into the realm of clinical trials and the regulatory processes involved in bringing medical devices to market.

Clinical trials are essential for evaluating the safety, efficacy, and performance of medical devices. These trials involve testing the device on human subjects to gather data on its effectiveness and potential risks. Students in the field of biomedical engineering should understand the different phases of clinical trials, from initial feasibility studies to large-scale trials involving thousands of participants. They should also be familiar with the ethical considerations and guidelines that govern these trials, ensuring the protection and well-being of study participants.

Regulatory approvals are another critical aspect of the medical device development journey. In this subchapter, students will gain insight into the regulatory bodies responsible for evaluating and approving medical devices, such as the Food and Drug Administration (FDA) in the United States. They will learn about the specific requirements and documentation necessary for submission, including preclinical data, clinical trial results, and manufacturing processes. Students will also discover the importance of post-market surveillance and adverse event reporting, as ongoing monitoring is crucial for ensuring the continued safety and efficacy of medical devices.

Furthermore, this subchapter will explore the challenges and complexities associated with regulatory approvals, such as the lengthy approval timelines and the need to balance innovation with patient safety. Students will gain an understanding of the role of regulatory professionals in navigating the regulatory landscape and ensuring compliance with the relevant standards and regulations.

By comprehending the intricacies of clinical trials and regulatory approvals, students in the field of biomedical engineering will be better equipped to contribute to the design and development of safe and effective medical devices. This knowledge will enable them to make informed decisions, adhere to regulatory requirements, and ultimately improve patient care through their future innovations in the field.

Chapter 8: Design for Sustainability and Environmental Impact

Sustainable Design Principles

In the field of biomedical engineering, sustainable design principles are becoming increasingly important in the development of medical devices. With the growing concern for the environment and the need for responsible resource management, it is crucial for students in this field to understand and implement sustainable design practices. This subchapter will explore the key principles of sustainable design and how they can be applied to create innovative and environmentally friendly medical devices.

One of the fundamental principles of sustainable design is to minimize the use of non-renewable resources. As students, it is essential to consider the life cycle of a medical device and its impact on the environment. By choosing materials that are renewable, recyclable, or biodegradable, you can reduce the device's carbon footprint and minimize waste generation. Additionally, the use of energy-efficient manufacturing processes and technologies can further enhance the sustainability of medical devices.

Another important principle is designing for longevity and durability. The goal is to create medical devices that have a longer lifespan, reducing the need for frequent replacements. By selecting high-quality materials and designing devices that can withstand wear and tear, you can minimize waste generation and conserve resources. Furthermore, incorporating modularity and upgradability into the design can allow for easy repairs and component replacements, extending the device's useful life.

Sustainable design also emphasizes the importance of energy efficiency. By utilizing energy-efficient components and technologies, medical devices can reduce their carbon footprint and conserve energy resources. For instance, incorporating power-saving features like sleep modes and automatic shut-offs can greatly enhance a device's energy efficiency. Moreover, exploring alternative energy sources, such as solar or kinetic energy, can further contribute to sustainable design in the field of biomedical engineering.

Lastly, sustainable design principles encourage the consideration of end-of-life options for medical devices. Designing devices with disassembly and recycling in mind can promote proper disposal and reduce environmental impact. Additionally, exploring opportunities for device reuse or repurposing can further minimize waste generation.

In conclusion, as students in the field of biomedical engineering, it is crucial to embrace sustainable design principles in the development of medical devices. By minimizing the use of non-renewable resources, designing for longevity and durability, prioritizing energy efficiency, and considering end-of-life options, you can contribute to a more environmentally friendly and responsible approach to medical device design. By incorporating these principles into your work, you can help shape a sustainable future for the biomedical engineering industry.

Lifecycle Assessment and Environmental Impact

In the field of Biomedical Engineering, it is crucial to consider the environmental impact of medical devices throughout their lifecycle. This subchapter will delve into the concept of Lifecycle Assessment (LCA) and its significance in designing sustainable medical devices. By understanding the environmental impact of these devices, students in Biomedical Engineering can contribute to the development of more eco-friendly and socially responsible healthcare solutions.

Lifecycle Assessment is a systematic approach that evaluates the environmental impacts of a product from its creation to its disposal. It encompasses the entire life cycle of a medical device, including the extraction of raw materials, manufacturing, distribution, use, and end-of-life management. By conducting an LCA, engineers can identify areas where improvements can be made to minimize environmental harm.

The initial phase of a medical device's lifecycle involves the extraction and processing of raw materials. Students must be aware of the environmental implications associated with mining and manufacturing processes. By seeking alternative materials with lower environmental footprints, such as biodegradable or recycled materials, students can contribute to reducing the impact of medical devices on the environment.

During the manufacturing process, energy consumption and waste generation are key factors to consider. Students should explore energy-efficient manufacturing techniques, such as using renewable energy sources and implementing recycling programs for waste materials. By minimizing waste and energy consumption, the environmental impact of medical device production can be significantly reduced.

Once a medical device is in use, its energy consumption, emissions, and potential for pollution must be considered. Students can focus on designing devices that are energy-efficient, have extended product lifetimes, and incorporate materials that can be recycled or easily disposed of without harm to the environment.

Finally, the end-of-life management of medical devices is crucial. Students should explore ways to promote recycling, reusing, or repurposing devices to minimize waste generation. Proper disposal techniques and the development of biodegradable materials should also be emphasized.

In conclusion, understanding the lifecycle assessment and environmental impact of medical devices is paramount for students in the field of Biomedical Engineering. By considering the entire lifecycle of a device, from raw material extraction to disposal, students can design more sustainable and environmentally friendly solutions. By incorporating renewable materials, reducing energy consumption, and promoting recycling and proper disposal, students can contribute to a greener future for healthcare.

Implementing Sustainable Practices in Medical Device Design

As students in the field of biomedical engineering, it is crucial to understand the importance of implementing sustainable practices in medical device design. With the growing concern for the environment and the need for sustainable solutions, it is essential for future engineers to develop devices that not only improve patient outcomes but also minimize their impact on the planet.

One key aspect of sustainable medical device design is the use of eco-friendly materials. Traditional medical devices often contain harmful substances or non-biodegradable materials, contributing to pollution and waste. By incorporating recyclable and biodegradable materials into the design process, engineers can create devices that are less harmful to the environment. For example, utilizing bioplastics or natural fibers can reduce the carbon footprint of medical devices while still maintaining their functionality and performance.

Another important consideration is energy efficiency. Medical devices consume a significant amount of energy, often leading to increased carbon emissions and higher energy costs. By designing devices with energy-saving features, such as low-power modes or efficient energy management systems, engineers can reduce the environmental impact while also extending the device's battery life. This not only benefits the planet but also improves the device's usability and convenience for patients.

Furthermore, incorporating the principles of the circular economy can greatly contribute to sustainable medical device design. Instead of following the traditional linear model of production and disposal, a circular economy focuses on reducing waste and reusing resources. This can be achieved by designing devices that are easily repairable, upgradable, and modular. By prolonging the lifespan of medical devices and enabling the replacement of specific components, engineers can reduce electronic waste and conserve valuable resources.

Additionally, it is crucial to consider the end-of-life management of medical devices. Proper disposal and recycling of devices play a significant role in reducing environmental impact. Designing devices with disassembly in mind, using materials that are easy to separate and recycle, and providing clear instructions for environmentally friendly disposal are essential steps to ensure sustainable practices are followed.

In conclusion, implementing sustainable practices in medical device design is of utmost importance for biomedical engineering students. By using eco-friendly materials, focusing on energy efficiency, adopting circular economy principles, and considering end-of-life management, future engineers can contribute to a greener and more sustainable healthcare industry. Embracing sustainability in medical device design not only benefits the environment but also improves patient care and promotes a healthier future for all.

Chapter 9: Innovation and Emerging Technologies in Medical Devices

Cutting-Edge Technologies in Medical Device Design

In the rapidly evolving field of biomedical engineering, advancements in technology have revolutionized medical device design. These cutting-edge technologies have paved the way for innovative solutions that are transforming the healthcare industry. In this subchapter, we will explore some of the most exciting and game-changing technologies that are shaping the future of medical devices.

One of the most prominent technologies in medical device design is 3D printing. This revolutionary technique allows engineers to create complex and customized medical devices with incredible precision. From patient-specific implants to prosthetics, 3D printing has opened up new possibilities for personalized healthcare. Students in biomedical engineering can explore the endless potential of 3D printing to design devices that are tailored to individual patients' needs, enhancing treatment outcomes and patient comfort.

Another cutting-edge technology that is revolutionizing medical device design is artificial intelligence (AI). AI algorithms can analyze vast amounts of medical data and provide valuable insights for diagnosis, treatment, and monitoring. Students can harness the power of AI to develop intelligent medical devices that can detect diseases at an early stage, alert healthcare providers in emergencies, and even assist in surgery. The integration of AI and medical devices has the potential to greatly improve patient care and outcomes.

Nanotechnology is another exciting area in medical device design. By manipulating materials at the nanoscale, biomedical engineers can create devices with enhanced properties such as increased sensitivity, durability, and biocompatibility. Nanodevices can be used for targeted drug delivery, non-invasive diagnostics, and tissue engineering. Students specializing in biomedical engineering can explore the fascinating world of nanotechnology to create devices that push the boundaries of medical science.

Furthermore, the Internet of Things (IoT) has emerged as a game-changer in medical device design. By connecting devices and collecting real-time data, IoT enables remote patient monitoring, predictive analytics, and improved communication between patients and healthcare providers. Students can leverage IoT technology to design devices that provide continuous monitoring, facilitate telemedicine, and improve patient engagement.

In conclusion, the field of biomedical engineering is constantly evolving, driven by cutting-edge technologies. 3D printing, artificial intelligence, nanotechnology, and the Internet of Things are just a few examples of the innovations shaping the future of medical devices. As students in this field, exploring these technologies and their applications can unlock a world of possibilities to design medical devices that improve patient outcomes, transform healthcare, and shape the future of medicine.

Artificial Intelligence and Machine Learning in Healthcare

The field of biomedical engineering is rapidly evolving, and one of the most exciting developments is the integration of Artificial Intelligence (AI) and Machine Learning (ML) in healthcare. These cutting-edge technologies have the potential to revolutionize the way we diagnose, treat, and manage diseases, ultimately improving patient outcomes and transforming the healthcare industry as a whole.

AI refers to the simulation of human intelligence in machines that are programmed to think and learn like humans. ML, on the other hand, is a subset of AI that focuses on using algorithms to enable machines to learn from data and make predictions or decisions without being explicitly programmed. When combined, AI and ML can unlock valuable insights from vast amounts of healthcare data, leading to more accurate diagnoses, personalized treatment plans, and enhanced patient care.

One of the primary applications of AI and ML in healthcare is in medical imaging. By training algorithms on large datasets of medical images, such as X-rays, CT scans, and MRIs, AI can help radiologists detect abnormalities with greater accuracy and efficiency. This not only speeds up the diagnosis process but also reduces the chances of misdiagnosis, improving patient outcomes.

Additionally, AI and ML algorithms can be used to analyze electronic health records (EHRs) and identify patterns that could indicate the onset of certain diseases or predict patient outcomes. By leveraging these insights, healthcare professionals can intervene earlier, providing timely and targeted interventions to prevent or manage disease progression.

Another area where AI and ML are making significant contributions is in drug discovery and development. Traditional drug discovery processes are time-consuming and costly, often taking years to bring a new drug to market. AI and ML algorithms can help researchers analyze vast amounts of biomedical data to identify potential drug candidates more efficiently. This accelerated process has the potential to bring life-saving treatments to patients faster and at a lower cost.

However, as exciting as AI and ML are, their integration into healthcare also raises important ethical considerations. Issues such as data privacy, algorithm bias, and accountability must be addressed to ensure that these technologies are used responsibly and ethically.

In conclusion, the integration of AI and ML in healthcare holds immense potential for the field of biomedical engineering. These technologies have the power to transform the way we approach diagnosis, treatment, and patient care. As students in this field, it is crucial to stay updated on the latest advancements in AI and ML, actively engage in research and development, and contribute to the ethical implementation of these technologies to shape the future of medical devices and healthcare.

Future Trends and Opportunities in Medical Device Design

As students pursuing a career in biomedical engineering, it is crucial to stay informed about the latest trends and opportunities in the field of medical device design. The healthcare industry is constantly evolving, and with advancements in technology, the potential for innovation in medical devices is expanding rapidly. This subchapter aims to provide an overview of the future trends and opportunities that await aspiring biomedical engineers in the realm of medical device design.

One significant trend that will shape the future of medical device design is the integration of artificial intelligence (AI) and machine learning (ML) algorithms. AI has the potential to revolutionize healthcare by improving diagnostics, enabling personalized medicine, and enhancing patient care. As a student, it is essential to familiarize yourself with AI and ML techniques, as they will play a vital role in designing medical devices that can analyze complex data, make accurate predictions, and assist healthcare professionals in making informed decisions.

Additionally, the rise of wearable technology presents exciting opportunities for biomedical engineering students. Wearable devices, such as smartwatches and fitness trackers, are becoming increasingly popular for monitoring vital signs, tracking physical activity, and even diagnosing certain conditions. Exploring the design and development of wearable medical devices can open doors to innovative solutions for remote patient monitoring, early detection of diseases, and improving overall patient outcomes.

Another area of growth in medical device design is the development of minimally invasive technologies. Traditional surgical procedures often come with risks and lengthy recovery times. However, advancements in robotics, imaging, and materials have paved the way for minimally invasive procedures that offer patients faster recovery, reduced scarring, and improved surgical outcomes. As a student, learning about the principles behind minimally invasive technologies and exploring their potential applications can provide you with a competitive edge in the field.

Furthermore, the increasing focus on personalized medicine presents opportunities for students interested in medical device design. Tailoring medical treatments to individual patients based on their genetic makeup and specific needs has the potential to revolutionize healthcare. Designing devices that enable precise drug delivery, genetic testing, or patient-specific implants can significantly impact the future of medicine and patient care.

In conclusion, the future of medical device design holds immense potential for biomedical engineering students. By understanding and embracing trends such as the integration of AI and ML, wearable technology, minimally invasive procedures, and personalized medicine, students can position themselves as innovators in the field. Embracing these opportunities will not only contribute to the advancement of healthcare but also enable students to make a meaningful impact on patient lives.

Chapter 10: Bringing Your Design to Market

Intellectual Property Protection

As aspiring biomedical engineers, it is crucial for us to understand the importance of intellectual property protection in the field of medical devices. In this subchapter, we will delve into the significance of safeguarding our innovative ideas and designs, and the various methods available to protect our intellectual property.

In the fast-paced world of biomedical engineering, where groundbreaking medical devices are constantly being developed, intellectual property protection plays a pivotal role. It ensures that the time, effort, and resources invested in creating these devices are duly recognized and rewarded. Moreover, it encourages innovation and drives the advancement of medical technology.

One of the most common methods of protecting intellectual property is through patents. Patents grant inventors the exclusive rights to their inventions for a certain period, preventing others from using, making, or selling their devices without permission. Obtaining a patent involves a thorough understanding of the patent application process, including conducting prior art searches and drafting detailed and accurate patent claims. It is essential for students in biomedical engineering to familiarize themselves with this process, as it will empower them to protect their ideas effectively.

Another method of intellectual property protection is through copyrights. Although copyrights primarily protect creative works such as literature, music, and art, they can also extend to software and other forms of intellectual property. Students involved in the development of medical device software should be aware of the potential for copyright protection and take the necessary steps to safeguard their creations.

Trademarks are another crucial aspect of intellectual property protection. While trademarks are primarily used to protect brand names, logos, and slogans, they can also be employed to protect unique designs or features of medical devices. By registering a trademark, students can prevent others from using similar marks, ensuring that their device stands out in the market.

Additionally, trade secrets can be utilized to protect intellectual property. Trade secrets involve keeping valuable information confidential, such as manufacturing processes or formulas. Students working on proprietary medical device designs should be aware of the importance of maintaining secrecy and implementing measures to protect their trade secrets.

In conclusion, intellectual property protection is of utmost importance in the field of biomedical engineering. By understanding and utilizing methods such as patents, copyrights, trademarks, and trade secrets, students can ensure that their innovative ideas and designs are safeguarded. This not only protects their hard work but also fosters a culture of innovation and drives the development of cutting-edge medical devices.

Business and Entrepreneurship in Medical Device Design

In the ever-evolving field of biomedical engineering, the intersection of business and entrepreneurship plays a crucial role in the development and success of medical device design. As students venturing into this dynamic field, it is important to understand the significance of business acumen and the entrepreneurial mindset in shaping the future of medical devices.

The integration of business principles into medical device design enables students to not only create innovative solutions but also bring them to the market successfully. Starting a medical device company requires a deep understanding of the industry landscape, regulatory processes, and market demands. By embracing the entrepreneurial mindset, students can navigate these challenges and transform their ideas into tangible products.

One of the key aspects of business and entrepreneurship in medical device design is the identification of unmet clinical needs. Students must closely observe healthcare practices, interact with healthcare professionals, and empathize with patients to identify gaps in existing medical technologies. This process of needs identification serves as the foundation for designing innovative medical devices that address these unmet needs effectively.

Furthermore, students must acquire knowledge in intellectual property protection, market analysis, and regulatory compliance to ensure the successful commercialization of their medical devices. Understanding the patenting process, conducting market research, and adhering to regulatory guidelines are essential steps in bringing medical devices from the design phase to the hands of end-users.

The business and entrepreneurial mindset also emphasize the importance of collaboration and networking. Students should actively engage with professionals in the medical device industry, participate in conferences and exhibitions, and seek mentorship from experienced entrepreneurs. Building a strong network not only opens doors to potential investors but also provides valuable insights and guidance throughout the product development and commercialization journey.

Moreover, students need to comprehend the financial aspects of medical device design. They must be able to develop a business plan, estimate costs, and secure funding to support the development, manufacturing, and marketing of their medical devices. Understanding the fundamentals of finance and investment is crucial in attracting investors and ensuring the sustainability of their ventures.

In conclusion, the subchapter on "Business and Entrepreneurship in Medical Device Design" emphasizes the vital role played by business principles and entrepreneurial thinking in the field of biomedical engineering. As students, it is imperative to cultivate business acumen, identify unmet clinical needs, navigate regulatory processes, build networks, and understand financial aspects to design successful medical devices. By embracing the entrepreneurial mindset, students can revolutionize healthcare with their innovative ideas and contribute to the betterment of society.

Funding and Commercialization Strategies

In the fast-paced world of biomedical engineering, where innovation is key to developing life-saving medical devices, securing funding and understanding commercialization strategies are crucial aspects of bringing a product to market. This subchapter will delve into the various funding options available to students and the strategies they can employ to successfully commercialize their designs.

1.   Funding   Options:
a. Grants: Students in the field of biomedical engineering have access to several grants and funding opportunities. National research agencies, foundations, and universities often provide grants specifically tailored to support innovative projects in medical device development. These grants can provide financial support for research, prototyping,   and   even   clinical   trials.
b. Competitions: Participating in innovation competitions can also provide funding opportunities. Many organizations and companies sponsor competitions that award cash prizes or investment opportunities to winning projects. These competitions not only provide funding but also valuable exposure and networking opportunities.
c. Crowdfunding: Another popular option for funding is crowdfunding platforms. Students can create campaigns to raise money from the general public for their medical device projects. Crowdfunding allows for direct engagement with potential customers and can provide valuable feedback during the development process.

2.   Commercialization   Strategies:
a. Intellectual Property Protection: Before commercializing a medical

device, students should prioritize intellectual property protection. This can be achieved through patents, trademarks, copyrights, or trade secrets. Protecting intellectual property ensures that the innovation remains unique and provides a competitive advantage in the market. b. Market Research: Conducting thorough market research is essential to understand the target audience, competition, and potential demand for the medical device. This information will help students refine their product design, pricing strategy, and marketing efforts. c. Collaborations: Partnering with industry experts, hospitals, or research institutions can provide invaluable support in terms of funding, expertise, and access to clinical trials. Collaborations can also enhance the credibility and marketability of the medical device. d. Regulatory Compliance: Students must be aware of the regulatory requirements and standards for medical devices. Understanding and complying with these regulations is crucial for commercialization success and ensuring patient safety.

By understanding the available funding options and implementing effective commercialization strategies, students in the field of biomedical engineering can bring their innovative medical device designs to life. This subchapter will equip students with the knowledge and tools necessary to navigate the complex landscape of funding and commercialization and ultimately contribute to the advancement of medical technology.

Chapter 11: Case Studies and Success Stories

Notable Medical Device Design Innovations

In the ever-evolving field of biomedical engineering, medical device design plays a crucial role in improving healthcare outcomes and enhancing patient well-being. From simple diagnostic tools to complex surgical equipment, medical device designers are constantly pushing the boundaries of innovation to create revolutionary solutions. This subchapter explores some of the most notable medical device design innovations that have revolutionized the healthcare landscape.

1.   Minimally Invasive Surgical Instruments: Traditionally, surgical procedures required large incisions, resulting in longer recovery times and increased risks. However, with the advent of minimally invasive surgical instruments, such as laparoscopes and endoscopes, surgeons can now perform complex procedures through tiny incisions. These devices offer improved patient comfort, reduced scarring, shorter hospital stays, and faster recovery times.

2.   Wearable Health Monitoring Devices: The emergence of wearable health monitoring devices has empowered individuals to actively manage their health. Devices like fitness trackers, smartwatches, and biosensors can measure vital signs, track physical activity, monitor sleep patterns, and provide real-time health feedback. These devices are invaluable in promoting healthy lifestyles, preventing diseases, and enabling personalized healthcare.

3.   Artificial   Organs:
The scarcity of donor organs for transplantation has led to the development of artificial organs. Devices such as artificial hearts, kidneys, and limbs have dramatically improved the quality of life for patients waiting for transplants. These innovative devices mimic the functions of natural organs, offering hope to those in need and reducing the dependency on donor organs.

4.   Prosthetic   Limbs:
Advancements in prosthetic limb design have revolutionized the lives of individuals with limb loss. Modern prosthetics incorporate advanced materials, robotics, and sensors to replicate natural movements and enhance functionality. Some prosthetic limbs even provide sensory feedback, allowing users to regain a sense of touch and proprioception.

5.   3D   Printing   in   Medical   Device   Manufacturing:
The integration of 3D printing technology in medical device manufacturing has opened up new possibilities. Customized implants, surgical guides, and prosthetic limbs can now be fabricated with precision, reducing surgery time and enhancing patient outcomes. 3D printing also allows for rapid prototyping, enabling designers to iterate and refine their ideas quickly.

These notable medical device design innovations highlight the transformative impact of biomedical engineering on healthcare. By continuously pushing the boundaries of innovation, biomedical engineers are improving patient outcomes, enhancing quality of life, and shaping the future of medicine. As students of biomedical engineering, understanding these innovative designs provides inspiration and motivation to contribute to this dynamic field, making a positive difference in the lives of countless individuals.

Inspiring Student-led Medical Device Projects

As a student in the field of biomedical engineering, you have the unique opportunity to make a significant impact on the healthcare industry through the design and development of medical devices. This subchapter, "Inspiring Student-led Medical Device Projects," aims to highlight some remarkable initiatives undertaken by students like yourself, showcasing their innovation and creativity in improving healthcare outcomes.

One inspiring example is the development of a low-cost prosthetic hand by a group of undergraduate students. Fueled by their passion to make prosthetics more accessible, they used 3D printing technology to design a functional and affordable hand. By leveraging their engineering skills and working closely with healthcare professionals, they were able to create a device that addressed the specific needs of amputees in their community.

Another student-led project worth mentioning is the creation of a wearable device that monitors and alerts individuals with chronic conditions about potential health risks. This device, developed by a team of graduate students, combines sensor technology and data analysis to provide real-time monitoring and personalized notifications. By empowering patients to take control of their health, this project has the potential to revolutionize how chronic diseases are managed.

In addition to these examples, there are numerous other student-led projects that have made significant contributions to the field of medical devices. Some students have developed innovative diagnostic tools, such as smartphone apps that can detect skin cancer or portable devices for early detection of infectious diseases. Others have focused on improving existing medical devices, such as designing more comfortable and functional prosthetics or developing advanced imaging systems for more accurate diagnoses.

These projects not only demonstrate the technical skills and knowledge of biomedical engineering students but also highlight their commitment to improving healthcare outcomes. By taking a hands-on approach to solving real-world problems, students are able to bridge the gap between theory and practice, making a tangible difference in the lives of patients.

If you are a student interested in pursuing a career in biomedical engineering, these inspiring student-led medical device projects serve as a testament to the power of innovation and collaboration. They showcase the potential impact you can have in the field and provide a glimpse into the exciting possibilities that lie ahead. By learning from these examples and embracing your creativity, you can contribute to the future of medical device design and help shape the healthcare industry for the better.

Lessons learned from Successful Medical Device Entrepreneurs

In the ever-evolving field of biomedical engineering, the role of medical device entrepreneurs cannot be understated. These visionary individuals have not only transformed the healthcare landscape but have also paved the way for aspiring students to make their mark in the industry. In this subchapter, we will delve into the invaluable lessons learned from successful medical device entrepreneurs, offering insights and inspiration to students pursuing a career in biomedical engineering.

1. Identify Unmet Needs: Successful medical device entrepreneurs possess a keen eye for identifying unmet needs in healthcare. They understand that innovation begins by recognizing the gaps and challenges existing in the current medical landscape. As students, it is crucial to develop a mindset that actively seeks out opportunities for innovation and improvement.

2. Think Outside the Box: The ability to think creatively and embrace unconventional approaches is a hallmark of successful medical device entrepreneurs. They understand that breakthroughs often come from challenging traditional norms and exploring uncharted territories. As students, it is important to foster a mindset of curiosity and open-mindedness, allowing us to break free from conventional thinking and push the boundaries of possibility.

3. Collaborate and Build a Strong Network: Building a strong network of like-minded individuals is instrumental in the success of medical device entrepreneurs. Collaboration with experts from various disciplines, such as clinicians, engineers, and business professionals,

can provide unique perspectives and invaluable support. As students, it is essential to actively seek out opportunities for collaboration and networking, both within and outside the academic setting.

4. Embrace Failure as a Stepping Stone: Failure is an inevitable part of the entrepreneurial journey. Successful medical device entrepreneurs embrace failure as a learning opportunity and a stepping stone towards success. Students should not fear failure but rather view it as a chance to grow, adapt, and refine their ideas.

5. Understand Regulatory and Reimbursement Landscape: Medical device entrepreneurs must navigate complex regulatory and reimbursement processes. Successful entrepreneurs have a solid understanding of these aspects and work closely with regulatory bodies to ensure compliance. As students, it is crucial to gain knowledge about these processes to develop devices that are not only innovative but also meet the necessary standards.

In conclusion, the lessons learned from successful medical device entrepreneurs provide valuable guidance for students pursuing a career in biomedical engineering. By identifying unmet needs, thinking creatively, collaborating, embracing failure, and understanding the regulatory landscape, students can develop the skills necessary to make a meaningful impact in the field. By combining these lessons with a passion for innovation, students can design the future of medical devices and contribute to improving healthcare worldwide.

Chapter 12: The Future of Medical Device Design

Predictions and Exciting Developments in the Field

As students in the field of biomedical engineering, it is crucial to stay updated on the latest predictions and exciting developments in the industry. The field of medical devices is constantly evolving, with new technologies and innovations emerging at a rapid pace. In this subchapter, we will explore some of the most promising predictions and developments that are shaping the future of biomedical engineering.

One of the most exciting developments in recent years is the rise of wearable medical devices. These devices have the potential to revolutionize healthcare by providing continuous monitoring and real-time data collection. Imagine a wristband that can track your heart rate, blood pressure, and oxygen levels throughout the day, providing valuable insights into your overall health. These wearables have the potential to detect early signs of diseases, allowing for timely intervention and improved patient outcomes.

Another exciting prediction in the field is the integration of artificial intelligence (AI) into medical devices. AI has the ability to analyze vast amounts of patient data and identify patterns that may go unnoticed by human physicians. This can lead to more accurate diagnoses and personalized treatment plans. For example, AI-powered imaging systems can analyze medical images, such as X-rays and MRIs, and detect abnormalities with high precision and efficiency.

Advancements in nanotechnology are also expected to have a profound impact on the field of biomedical engineering. Nanodevices, with their tiny size and unique properties, can be used for targeted drug delivery, tissue engineering, and even monitoring cellular activities at the molecular level. These nanodevices have the potential to greatly enhance the effectiveness of treatments and reduce side effects.

Furthermore, the field of regenerative medicine holds great promise for the future. Scientists are working on developing techniques to repair or replace damaged tissues and organs using stem cells and tissue engineering. This could potentially revolutionize the treatment of various diseases and injuries, such as spinal cord injuries, heart diseases, and organ failure.

In conclusion, the future of biomedical engineering is full of exciting possibilities. Wearable devices, AI integration, nanotechnology, and regenerative medicine are just a few areas that are expected to transform the field. As students, it is essential to stay informed about these developments and be prepared to contribute to the advancement of medical devices. Embracing these predictions and exciting developments will allow us to design innovative solutions that improve the quality of healthcare and save lives.

The Role of Students in Shaping the Future of Healthcare Technology

As students in the field of biomedical engineering, you have a unique opportunity to shape the future of healthcare technology. In today's rapidly advancing world, medical devices are becoming increasingly sophisticated, offering innovative solutions to complex healthcare challenges. Your knowledge, creativity, and passion can contribute significantly to this transformative process.

One of the key roles students play in shaping the future of healthcare technology is through research and development. As students, you have the advantage of fresh perspectives and an eagerness to explore uncharted territories. By conducting research in areas such as medical imaging, diagnostics, prosthetics, or telemedicine, you can contribute to the development of new and improved devices that enhance patient care and outcomes.

Moreover, students can actively participate in the design and prototyping of medical devices. With advancements in technology and access to resources, you can bring your ideas to life and create prototypes that address specific healthcare needs. By collaborating with healthcare professionals and other stakeholders, you can ensure that your designs are practical, user-friendly, and aligned with the requirements of the healthcare industry.

Students also have a crucial role to play in ensuring the safety and efficacy of medical devices. Through rigorous testing and analysis, you can identify potential risks and address them before the devices are deployed in real-world scenarios. By adhering to ethical guidelines and regulatory standards, you can contribute to the development of safe and reliable medical devices.

Furthermore, as students, you have the power to advocate for healthcare technology and promote its benefits. By raising awareness about the potential of medical devices, you can inspire others to pursue careers in biomedical engineering and contribute to the field. Additionally, you can engage in discussions and debates surrounding healthcare policies and regulations to ensure that technology is utilized to its fullest potential in improving patient care.

In conclusion, as students in the field of biomedical engineering, your role in shaping the future of healthcare technology is indispensable. Through research, design, testing, and advocacy, you have the power to drive innovation and revolutionize patient care. Embrace this unique opportunity and let your passion for improving healthcare through technology guide you on this exciting journey.

Conclusion: Empowering Students to Design the Future of Medical Devices

Congratulations, future biomedical engineers! You have embarked on a journey that holds immense potential to revolutionize the healthcare industry. Throughout this book, "Designing the Future: A Student's Approach to Medical Devices," we have explored the exciting world of medical device design and the crucial role that you, as students, play in shaping its future.

From the very beginning, it became evident that the field of biomedical engineering is not just about inventing new gadgets; it is about improving and saving lives. As students, you possess a unique perspective that can bring fresh ideas and innovative solutions to the table. You have the power to challenge the status quo and push boundaries, ultimately transforming the way we approach healthcare.

Throughout the chapters, we have delved into the fundamental principles of medical device design, taking into account the needs of patients, healthcare professionals, and the broader society. We have explored the importance of conducting thorough research, collaborating with interdisciplinary teams, and considering ethical implications. By embracing these concepts, you are well-equipped to design medical devices that are not only effective but also safe, user-friendly, and accessible to all.

Furthermore, we have emphasized the significance of empathetic design. By putting yourself in the shoes of patients and understanding their experiences, you can create devices that truly address their needs and improve their quality of life. This human-centered approach is at the core of successful medical device design, and you, as students, have the creativity and empathy to excel in this aspect.

As the book comes to a close, we want to underscore the importance of your role in shaping the future of medical devices. The challenges faced by the healthcare industry are vast, ranging from the aging population to the increasing burden of chronic diseases. However, we firmly believe that with your passion, dedication, and skills, these challenges can be transformed into opportunities for innovation.

Remember, the future of medical devices is not just about inventing groundbreaking technologies, but also about ensuring their seamless integration into healthcare systems. This includes considerations of regulatory frameworks, manufacturing processes, and sustainability. You have the power to drive change on every level, making the healthcare landscape more efficient, equitable, and patient-centered.

So, dear students, embrace your potential as biomedical engineers and seize the opportunity to design the future of medical devices. Continue to learn, collaborate, and dare to dream big. Your ideas and contributions are invaluable, and together, we can create a healthier, brighter future for all.

Thank you for joining us on this journey, and we wish you all the success in your endeavors as you design the future of medical devices.

Stay inspired, stay curious, and keep innovating!

www.ingramcontent.com/pod-product-compliance
Lightning Source LLC
Chambersburg PA
CBHW050600160726
48003CB00002B/981